PAINTS

Henry Pluckrose

Photography: Chris Fairclough

FRANKLIN WATTS
London / New York / Sydney / Toronto

Copyright © 1987 Franklin Watts

Franklin Watts Inc
387 Park Avenue South
New York
N.Y. 10016

US ISBN: 0 531 10471 0
Library of Congress Catalog
Card Number: 87 50906

Design: Edward Kinsey

Editor: Jenny Wood

Printed in Belgium

The author wishes to record his thanks in the preparation of this book to: Hilary Devonshire for her help in the preparation of material; Christopher Fairclough for the excellence of his step-by-step photographs; Chester Fisher, Franklin Watts Ltd, for his advice and guidance; Roz Sullivan for her careful preparation of the typescript; and all the young people whose work he drew upon to illustrate this book.

Contents

This book describes activities which use the following:

Acrylic medium
Cardboard
Charcoal
Clear furniture polish
Cold water paste
Drinking straws
Gravel (or fine sand)
India ink
Jars for water
Knit glove
Large water bowl or tray
Printing roller (or household paint roller)
Liquid detergent
Marbling colors
Mixing trays
Paint – must be water-based – can be bought as:
  (a) powder color;
  (b) tempera blocks;
  (c) small bricks or tubes of artists' water color;
  (d) tubes of acrylic color;
  (e) finger paints;
  (f) poster colors;
  (g) texture colors;
  (h) tempera paste.
  Make sure you buy at least these five basic colors: red, yellow, blue, black and white.
Paintbrushes (varying sizes and types)

Palette knife
Paper – cartridge, sugar or construction paper
Paper clips
Pencil
Saucer or plate
Scissors
Scrap material – e.g. pieces of bark, braid, buttons, card, cotton, fur, hair, lace, leather, newspaper and paper scraps of all kinds, paper plates, paper tissues, plastic, raffia, ribbon, sacking, sequins, shells, small stones, sponges, twigs, small pieces of wood, wool
Tablespoon
Texture paste (for use with acrylic colors)
Water

This book has been prepared to encourage you to experiment with color. The activities and ideas outlined in the pages that follow will help you to discover how particular types of paints behave, the many different ways in which color can be applied and the kinds of surfaces which are suitable for picture-making.

There is no one way to paint a picture. All artists experiment with color, shape, pattern and texture. Each uses materials in a personal way.

Once you have learned how materials behave, experiment for yourself. To be creative is not just following instructions. The creative person takes an idea and turns it into something which is his or her own.

Some hints

Before you begin, remember that art activities can be messy!

If you are using this book at home, take all the precautions you would take at school. Cover the table with newspaper and the floor around it with an old cotton sheet or a piece of plastic (e.g. a large plastic garbage bag).

Cover yourself too. An old shirt or blouse, particularly if it is too big, will give you excellent protection from head to toe.

1 Water-based paints are sold in many different packagings. Always read the instructions before you begin.

Your paint

How you use paint will depend upon the type of paint you have. All of the paints mentioned in this book are water-based. This means that, to work with them, you will need water.

(a) Powder color is in powder form and has to be mixed with a wetting agent (e.g. water, liquid detergent) before it can be used. Always add water to powder – never powder to water.

(b) Tempera blocks are solid blocks of color. The surface of each block needs to be thoroughly dampened before use.

(c) Artists' water color is sold in small bricks and in tubes. It can be expensive.

(d) Acrylic color is a thick creamy paint, usually sold in large tubes. Acrylic colors have a plastic base and can be thinned with water. Particular care has to be taken when using acrylic paint as its base is also a strong adhesive (glue). When not being used (even for a short period), paintbrushes must be cleaned or kept in water. If this is not done, the brush heads will dry solid. (You can make your own acrylic colors by mixing powder color with acrylic medium – see pages 31–32.)

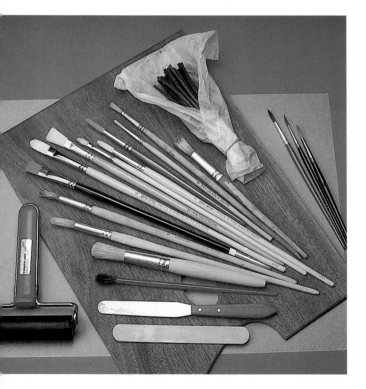

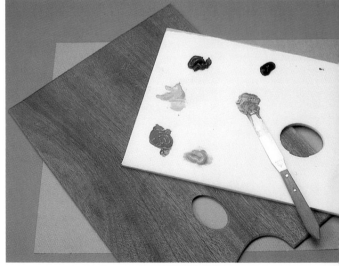

2 Color can be applied in many ways. Notice the range of brush heads.

3 Palettes like these are useful when working with acrylic color.

4 Muffin and cake pans make excellent palettes.

(e) Finger paint colors have been specially prepared for painting with the hands. (You can make your own finger paint mix by mixing water paste and powder color – see page 47.)

(f) You can use other kinds of water-based paints (e.g. poster colors, texture colors, tempera paste). Oil-based paints are not suitable for the activities described in this book, except for those on pages 28–30.

Your brushes

Brushes are sold by size. The lower the number (e.g. 2), the smaller the brush. Try to build up a collection of brushes in sizes 2, 4, 6, 8, 10 and 12.

5 Artists' water color, palette and brushes.

Some brushes are made with a "flat" tip, while others are shaped to a point. Build up a range of brush types.

Brushes are made in different materials. Some (like squirrel hair and sable) are very soft. Others (like hog hair) are hard. Some are made with synthetic materials.

Remember – no one brush is suitable for every activity!

6 Build up a stock of paper of different types and colors. Try to select a paper for your painting which links to the subject.

Your paper

Pictures can be painted on almost any kind of paper. The type of paper chosen will depend upon the type of paint being used. For example, thin tissue paper would not be a very suitable background for heavy layers of acrylic paint.

Finger painting

Finger paintings can be worked with thick pre-mixed paints on almost any type of paper. The most important thing to remember when using finger paint is to work quickly. You can draw into the paint only while it is wet.

Prints can also be taken from finger paintings.

1 Spread a quantity of finger paint into a clean tray.

2 While the paint is wet, draw a picture into it with the tip of your finger.

3 Place a clean sheet of paper over your picture (the paint must still be wet) and smooth it down with the backs of your hands. Do not press too hard.

4 Pull the paper away carefully. A print or transfer of the picture will now be on the paper. How does the print differ from the original?

5 This picture, if it is still wet, can also be used to give a transfer. Simply place a piece of paper over it, smooth down and peel off.

6 The finger painting (on white paper) and its transfer picture.

7 *The Scarecrow*
Finger paints can also
be used to make
pictures by painting
with the fingers
directly on to paper.

Finger paint mix can also be used to make attractively patterned paper. When dry, a sheet of color-combed paper makes an excellent book cover. Before using a sheet of combed paper to cover a book, you will need to "x" the color by painting a thin coat of shellac over the pattern. Make sure the paint is dry first!

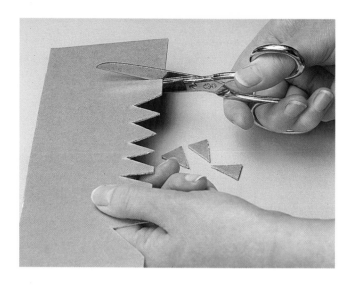

1 Make a comb from cardboard by cutting notches along one edge.

2 Spread finger paint mix into a tray. Draw in it with your comb.

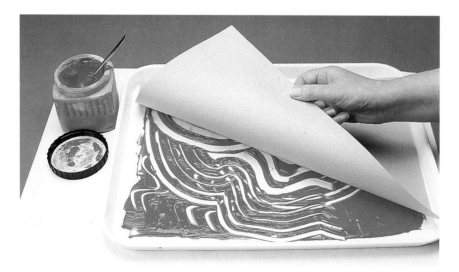

3 While the paint is still wet, place a sheet of paper over the design. Smooth it down lightly with the backs of your hands.

4 The combed sheet.

Applying color with wood

Experiment to discover how different materials make different kinds of marks in wet paint. Would the soft edge of a feather, for example, give the same sort of line as the piece of wood that has been used to draw the knight shown here?

1 Put two spoonfuls of thick color on to a sheet of paper.

2 Spread the color evenly over the paper's surface.

3 While the paint is still wet, draw into it with a small piece of wood.

4 *Knight* Why is it important to choose a paper which is a contrasting color to the paint?

Another way of applying color is with a roller. The type shown here is designed for linoleum blocks but a small household paint roller would do just as well. The roller is used to texture and color the paper. The texture provides the painting background.

1 (Above) Squeeze some color into a tray. Soak the roller with color and roll it back and forth across the paper.

2 (Below) When the textured paper is dry, paint a design on to it. The design could be made up of simple lines or of blocks of color.

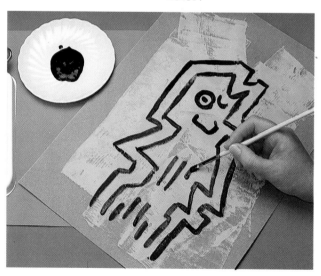

3 *The Old Chair In Sunlight* Here the textured sheet has been used to provide the background for a silhouette.

4 *At the Concert*
Here the silhouettes
are in two colors.

5 *The Strange Bird*
More than one color
can be used for the
textured background.

6 *The Park*
Here the roller has
been used to paint
most of the picture –
in shades of green.

Sort your scrap material (see page 4) into several piles. Into one, put all the soft materials (e.g. scraps of wool, fabric, sacking) and into the other all the hard materials (e.g. wood, bark, twigs). Divide each pile into those materials which have a heavy surface texture (like bark) and those which have very little (like smooth plastic). Experiment with the materials you have collected to see what kind of "mark" they make when used to apply color. Could you make a picture using only soft materials to apply color? Could you make a picture using only hard materials?

1 A simple way to soak scrap material with color is to make a color pad with a piece of sponge. Place the sponge in a saucer and pour paint on to it. Soak the scraps with color by pressing them into the paint-soaked sponge.

2 *Wild Cat* This picture was painted using a small scrap of sponge to apply the color.

For this activity you will need an old knitted glove. It is easy to apply color cleanly and evenly with a glove. Try wrapping other kinds of textured material tightly around one hand. Charge your covered hand with color and use it to produce a pattern. Vary your pattern by working in different colors.

1 Roll out some color into a tray.

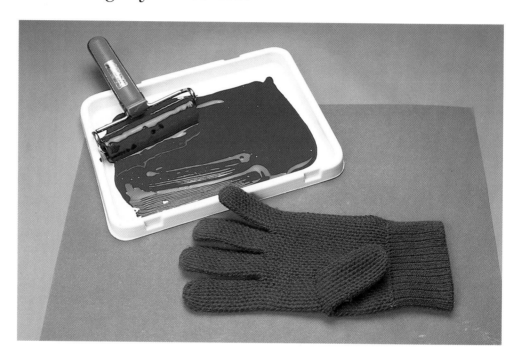

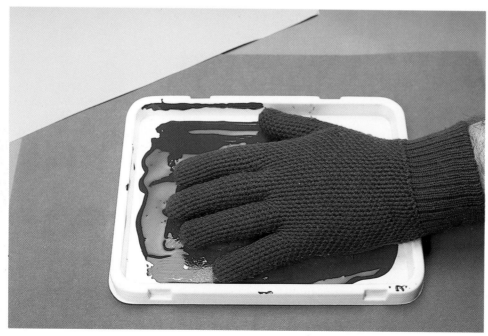

2 Charge the glove with color.

3 Apply the paint-soaked glove to paper.

4 *The Old Glove*
Notice how the paint echoes the texture of the material.

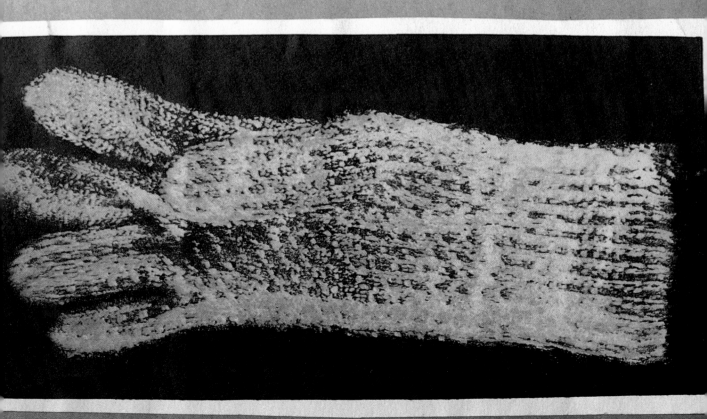

Applying color with a straw

Pictures can be made by applying color in tiny dots or spots. A simple way to make a dot picture is to apply the color with a drinking straw.

1 Place some thick paint into a palette. Here tempera paste has been squeezed on to paper plates – one plate and one straw for each color. The picture is built up by using the tip of the straw to apply color.

2 *Church Window*
A dot picture painted entirely with drinking straws.

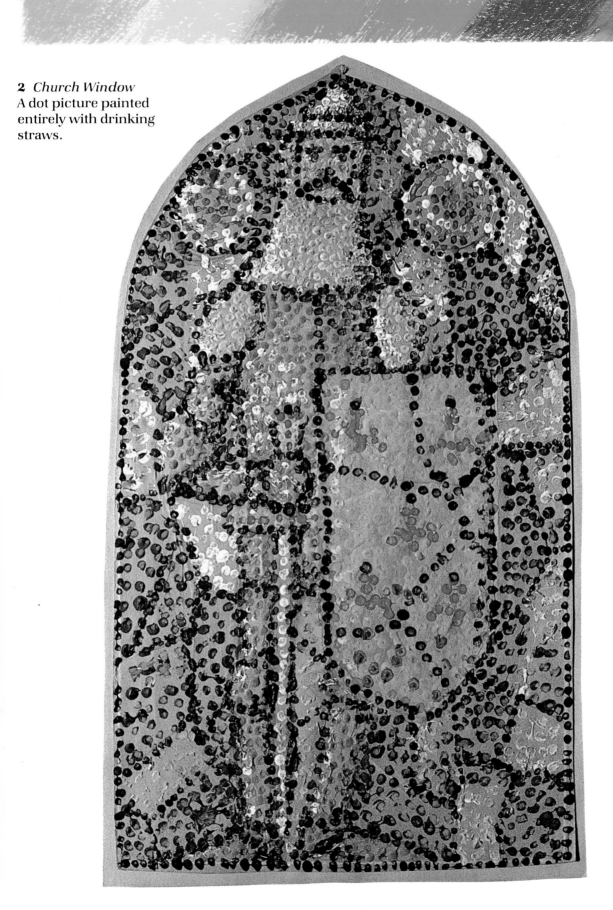

Applying color with card

As we have seen, color can be applied with the surface of a piece of cardboard. It can also be applied with its thin edge.

1 Take a piece of cardboard (a postcard will do well) and fold it against itself to make an interesting shape. Use a paper clip to hold the edges together.

2 Soak the edges with color and use them to make a pattern or picture.

3 *Trees*
You could also try to
make a picture by
applying color with the
edge of an unfolded
piece of card.

It's not even necessary always to use brushes, rollers or other materials to apply color. It can be fun simply to apply color by dripping paint directly from the tube. This is sometimes called "action painting".

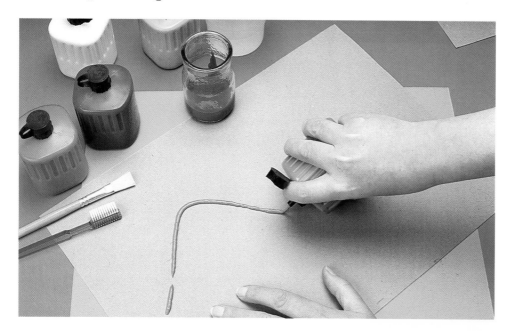

1 Squeezing out the color.

2 Adding drips of color.

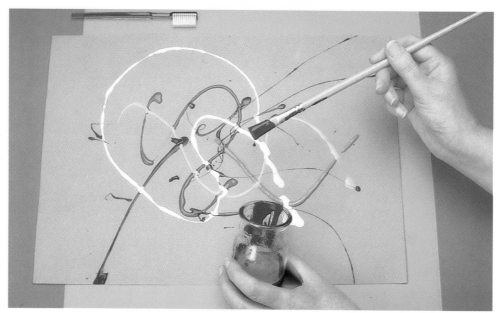

3 *Into Space* A splatter and drip picture.

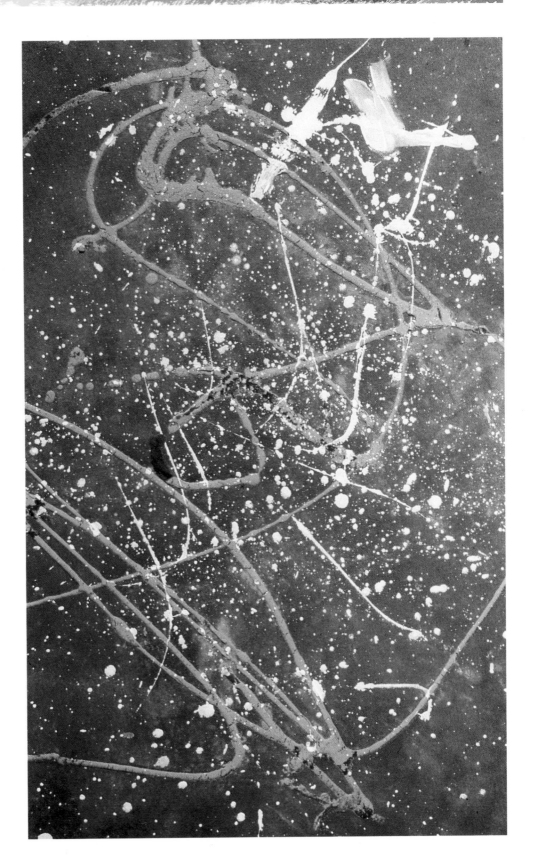

Marbling colors are special oil-based colors. They are used for making patterned papers.

When dry, marbled papers can be used for book covers, shelf and drawer liners, in model-making and even for writing paper.

1 Make sure all the materials are ready and close at hand before you begin. You will need a bowl of cold water, marbling colors, some straws, and sheets of paper on which to work.

2 Using the tip of a straw, drip a few spots of marbling color on to the water. Use more than one color.

3 Stir the surface of the water to make sure that the color spreads.

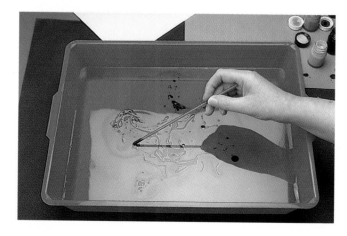

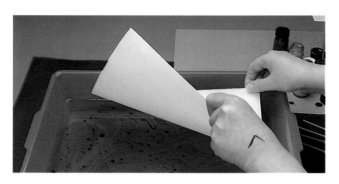

4 Drop a sheet of paper on to the water and let it float for a few moments.

5 Remove the paper and let dry.

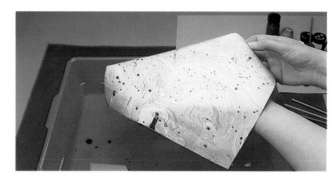

6 A marbled sheet.

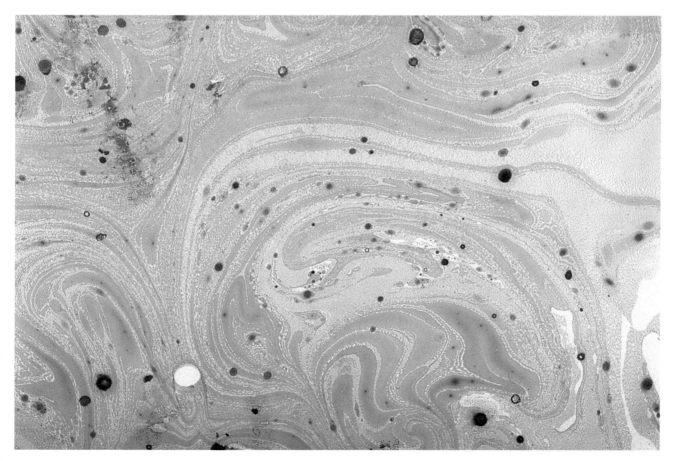

7 *My Friend* Painted
on marbled paper.

Acrylic color is a very thick color and can be applied with brush, roller or palette knife. But the plastic base can be thinned with cold water, so the color can also be applied in pale color washes.

Like all water-based colors, acrylic colors can be mixed with each other on a palette to give new colors and tones.

When making a picture with a thick color, it is wise to work on heavy paper or thin cardboard.

1 Put one heaped tablespoonful of powder color into a jar.

2 Slowly add water, stirring the powder until it looks like thick cream.

3 Add acrylic medium to the paint and water until the mixture is stiff and difficult to stir.

4 Tip a small quantity of color on to a palette, and paint – perhaps using a palette knife.

5 (Below) *St Paul's* A knife painting in blue and white acrylic colors. Notice how the color remains textured when dry.

6 *Strange Face*
A painting on
cardboard in acrylic
colors and using a
palette knife.

Because acrylic color is also an adhesive, other materials can be added to it to increase its texture. For example, you can get interesting effects by stirring fine sand (or even gravel) into the color before it is applied to the paper.

Some art suppliers sell extenders and texture pastes to use with acrylic colors.

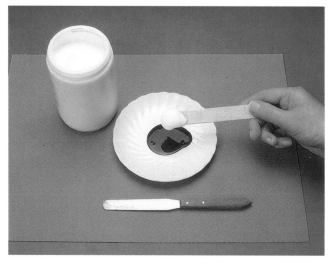

1 (Above) Adding texturing medium to acrylic paint.

2 (Left) Mixing medium and paint together with a palette knife.

3 Painting with a palette knife.

4 (Opposite) *Abstract* A texture painting.

Collage with acrylic colors

A collage is a word used to describe a picture that is made up of scraps of material.

Acrylic colors are also adhesive so if buttons, fabric and paper scraps, sequins, shells, sand or small stones are laid into the paint while it is wet the pieces will stick firmly to it and become part of the finished picture. The scraps can be given additional texture by painting over them with acrylic color.

1 Paint a picture in acrylic color.

2 While the paint is still wet, add details using scrap material.

4 (Opposite) *The Clown* How many different materials have been used here?

3 Here the collar is being decorated with sequins.

5 *At Sea* Here the picture is made from torn newspaper, a drinking straw and acrylic paint.

6 *A Welsh Lion Called Jones* This picture, painted by a seven-year-old, used cut and torn paper over a thin layer of acrylic color.

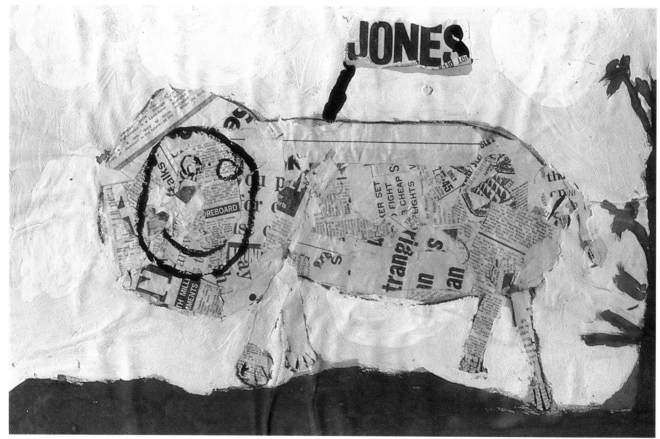

Marbled paper (see pages 28–30)
can also be used to make collages.
All you need is a selection of
marbled paper in different colors.
Details can be added in paint.

1 Here a picture of a
landscape is being
worked. The sheets of
marbled paper have
been cut into
mountain, hill and tree
shapes and glued on to
a sheet of red paper.
Fine details are being
added in black paint.

2 *Landscape* A collage
in marbled paper.

Using wet paper

Most of the pictures described in this book have been worked on dry paper. Working on wet paper produces quite a different result. You will find that when paints run together, unexpected blends of color occur.

1 Thoroughly wet the paper with clean water.

2 Apply washes of color.

3 *Welsh Hillside* A
painting on wet paper
by a nine-year-old.

Bubble patterns

This is another way of patterning paper in an unusual way. The papers produced can be used for book covers and for collage.

1 Squeeze a small quantity of liquid detergent into a jar.

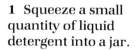

2 Add a spoonful of wet color (or India ink) and mix them together.

3 Blow the mixture through a drinking straw until bubbles rise over the edge of the jar.

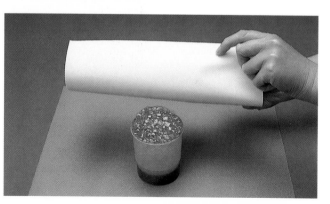

4 Lay a sheet of paper over the bubbles.

5 The bubble pattern will transfer to the paper.

6 Repeat the process until the whole page is decorated. You could make a pattern in several colors.

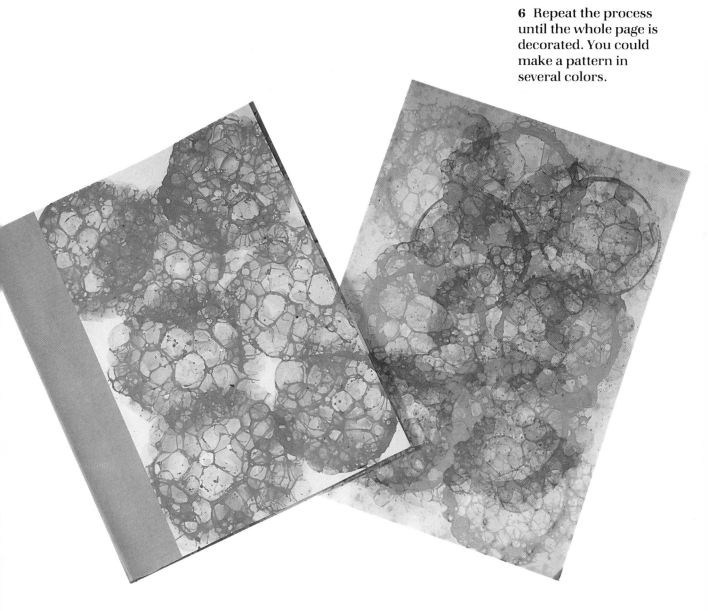

India ink is waterproof. Water-based paint is not. This difference means they will not blend together directly and can be used to produce fascinating pictures.

1 Using water-based paint, draw a picture in lines (i.e. do not fill with color). Leave to dry.

2 Brush waterproof India ink over the painting. Apply the ink thickly but try not to disturb the paint. Leave to dry.

3 Now soak the whole picture in water. With your fingertips, rub the areas you have painted. Because the paint dissolves in water, it will float away, taking with it the covering of India ink.

4 When all the paint has been removed, carefully lift the picture from the water. Leave to dry.

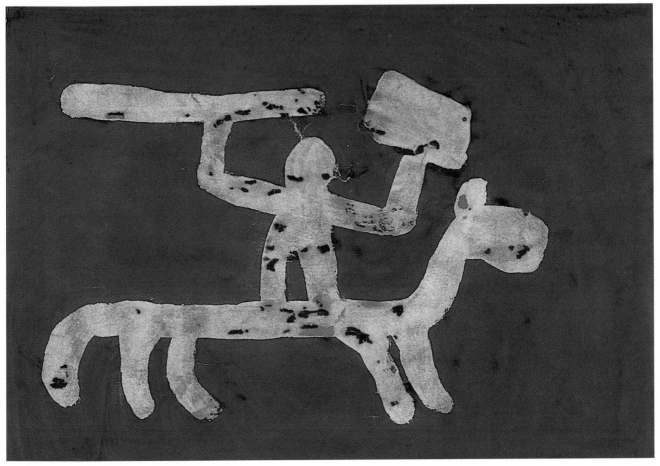

Throughout this book you have been experimenting with paints of different kinds. In doing this you have learned some of the ways in which paints behave. Now try some ideas of your own. Here are some suggestions to help you.

1 Could you make a picture which includes two or more of the ideas contained in this book? For example, could you use marbled paper for a picture worked in acrylic applied with a palette knife?

2 Could you make a picture applying dry powder color directly on to wet paper? What happens to the color? How easy is it to control?

3 Experiment by using a technique of your choice on different types of paper. For example, the effect achieved by applying watercolor washes to lightweight paper is very different to that achieved by applying watercolor washes to construction paper.

4 Try different types of color for marbling. What is the effect of laying yellow pastel paper in a tray of orange marbling colors?

5 Experiment by using unusual painting surfaces. For example, you could try making pictures on tissue paper, corrugated cardboard, crêpe paper or newsprint. How does the surface of the paper affect your picture, in the way you apply color and in the paints you use?

6 Use acrylic paint or thick poster color to decorate and pattern pebbles and smooth stones. Find some oval-shaped pebbles and decorate them with strange faces ... or turn them into weird animals and birds. Acrylic colors are excellent for painting on stone. If you use poster color, protect the design by applying a final coat of thin varnish. Do this when the paint has dried hard!

7 Look in an encyclopedia for information about Newton's color wheel. Make a wheel of your own. What does this teach you about color?

8 Try painting pictures and patterns using a restricted range of colors e.g. blue and white; white and black; red and yellow; blue and yellow; red and blue.

Art supply stores carry a range of materials such as brushes, papers and paints. Some art departments in which you can buy specialist materials like acrylic medium and texture paste.

A useful guide to stationery stores and art supply stores can be found in the *Yellow Pages*. (Look under Artists' Materials and Supplies and Graphic Arts materials'.)

Specialist materials (or materials in large quantities) can be purchased through a school supplier.

Acrylic colours and medium

Many of the major art supply stores manufacture their own range of acrylic colors.

Acrylic medium (for mixing with powder colour to produce an acrylic paint) can be purchased (at very little cost) in small jars. If the medium is being used for a group activity it is economic to buy it in quart or gallon containers.

Most acrylic medium dries to a hard skin that cannot be dissolved in water. However, some acrylic paints that will wash off clothing has been introduced.

Acrylic additives

Additives are used to give texture to acrylic color. They are sold under brand names. Each additive will be made to a slightly different recipe. Always read the instructions before you begin.

Finger paint

Finger paint can be made by mixing water paste and powder color. Add water to two tablespoonsfuls of powder color until a thick cream is formed. In a separate jar, mix two tablespoonsful of cold water paste with water until a thick smooth cream is formed. Now pour the color on to the paste and stir thoroughly. If the mixture is too thick, add a little water. If it is too thin, stir in dry color to thicken the mixture.

Marbling color

It is best to use specially prepared marbling color, but most oil-based can be used to make marbled patterns.

PRINTED IN BELGIUM BY

INTERNATIONAL BOOK PRODUCTION